Start with SCIENCE

Magnets

BY
CHARIS MATHER

BookLife
PUBLISHING

©2023
BookLife Publishing Ltd.
King's Lynn, Norfolk
PE30 4LS, UK

A catalogue record for this book is available from the British Library.

ISBN: 978-1-80155-829-7
ISBN: 978-1-80155-862-4

Written by:
Charis Mather

Edited by:
William Anthony

Designed by:
Drue Rintoul

PHOTO CREDITS

Cover – Anatoliy Karlyuk, Kokhanchikov, Matthew Dixon, AndreyProekt, Mauro Rodrigues, max dallocco. 4–5 – Tatiana Gordievskaia, Zen Wuak. 6–7 – Pat_Hastings, Ronald Rampsch, MicroOne. 8–9 – Hayati Kayhan, New Africa. 10–11 – noisuk Photo, titov dmitriy. 12–13 – kak2s, Mopic. 14–15 – Yusev, Rawpixel.com. 16–17 – xpixel, Diego loppolo. 18–19 – Bjwair, goodbishop. 20–21 – Andrus Ciprian, MR-R, Tarzhanova, Chinnapong. 22–23 – Maples Images, Rob Hyrons, BW Folsom, Yeti studio, Gerisima, Gerisima, Matveev Aleksandr, Hayati Kayhan, Atiwat Witthayanurut, Turbojet. All images are courtesy of Shutterstock.com, unless otherwise specified. With thanks to Getty Images, Thinkstock Photo and iStockphoto.

Contents

PAGE 4   Think like a Scientist

PAGE 6   Magnets

PAGE 8   Materials

PAGE 10   North and South Poles

PAGE 12   The Earth

PAGE 14   Compasses

PAGE 16   Magnet Patterns

PAGE 18   Animals

PAGE 20   In the Real World

PAGE 22   Be a Scientist

PAGE 24   Glossary and Index

Words that look like this can be found in the glossary on page 24.

Think LIKE A Scientist

There are lots of things in the world that we might not understand the first time we see them. Scientists are people who learn about these things.

When scientists make a guess about how something works, they test it to see if they are right. They pay close attention to see if any changes happen because of their tests.

Scientists share what they learn with others.

Magnets

Magnets

Magnets can be different shapes and sizes.

Scientists have learnt a lot about magnets. We use them every day, but what are they? A magnet is a rock or piece of metal that can pull some types of things towards it.

Magnets pull things towards them using a power called force. Every magnet has an *invisible* area of force around itself called a field. A magnetic field can pull some objects in.

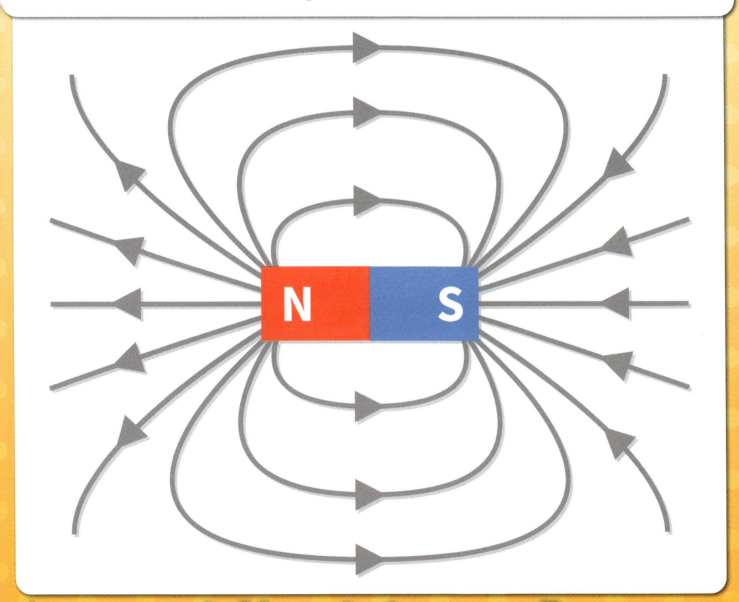

A magnetic field can be big or small.

Materials

A magnet's pull can make it stick to some things. Not everything is affected by a magnet's field, though. Wood, paper, plastic and many other materials are not magnetic. They cannot be pulled by magnets.

These objects are not magnetic.

A material that is affected by magnets is metal. Magnets only stick to some types of metal. This is why magnets can stick to some fridges and radiators, but not to drinks cans or some necklaces.

Is your fridge magnetic?

North AND South Poles

Magnets have two ends, called poles. One pole is called the north pole and one is called the south pole. When there are two magnets, the opposite poles are pulled together.

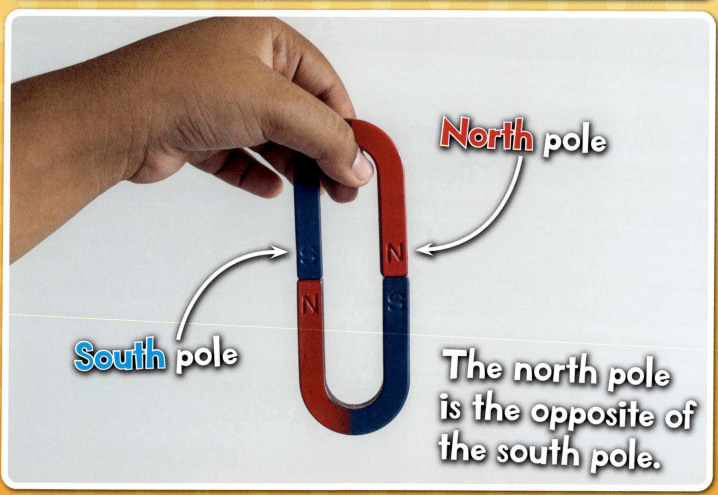

North pole

South pole

The north pole is the opposite of the south pole.

If you try putting two north poles or two south poles together, the magnets will push away from each other.

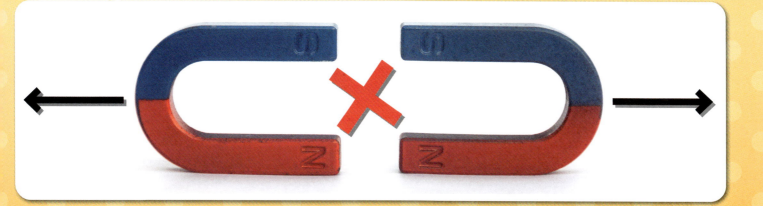

Magnets are a bit like building blocks because only the opposite ends stick together.

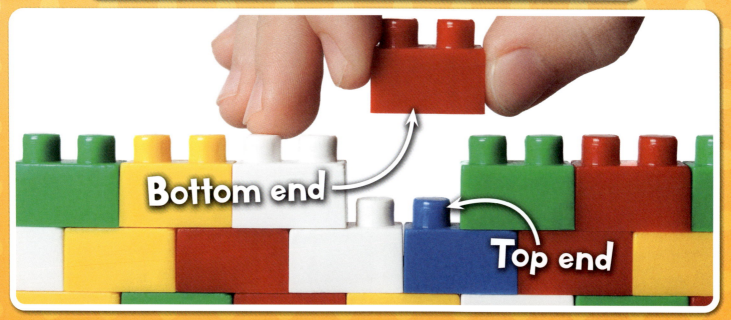

Bottom end

Top end

THE Earth

The Earth also has a North Pole and a South Pole. This is because the Earth acts like a giant magnet. People and animals can use the Earth's poles to tell direction.

This model of Earth helps us see where the poles are.

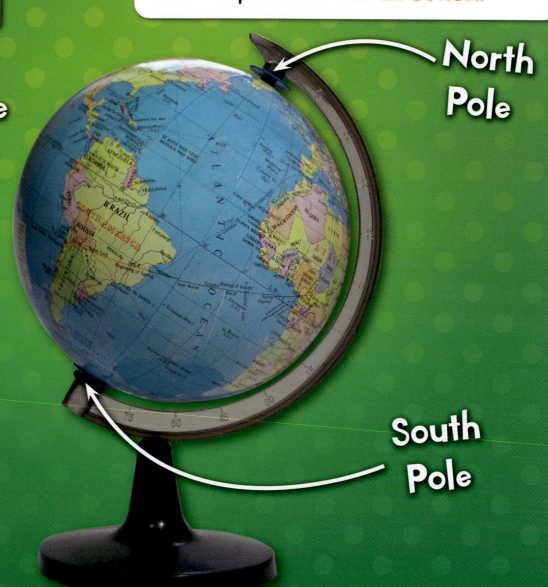

North Pole

South Pole

The Earth has its own magnetic field, just like smaller magnets. The magnetic field of the Earth is much larger than the magnetic field of the magnets in your house.

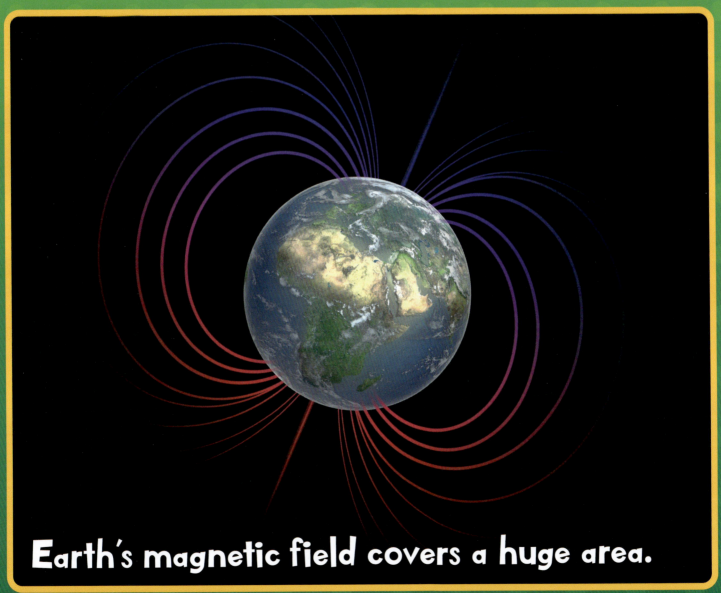

Earth's magnetic field covers a huge area.

Compasses

Compasses are tools that we use to find direction. Compasses have a thin piece of magnetic metal in the middle, called a needle. The needle points to the Earth's North Pole.

If you held a compass in your hand and slowly turned in different directions, the needle would still point north. That is because the magnet inside lines up with the Earth's poles.

Magnet Patterns

We can tell where a magnetic field is by doing a test. We can use a magnet and lots of tiny bits of metal called iron filings. Iron filings are attracted to magnets.

Iron filings look like a dark powder.

When iron filings are spread out on a piece of paper above a magnet, they make an unusual pattern. The iron filings group into lines that follow the magnetic field. The lines bend around the magnet.

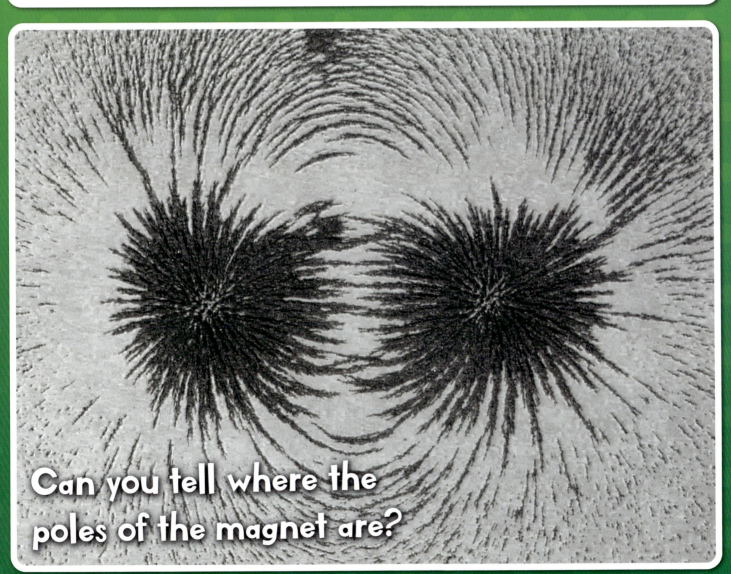

Can you tell where the poles of the magnet are?

Animals

Some animals, such as birds and fish, travel a long way at different times of the year. Animals do not have maps or compasses like we do, so how do they know where they are going?

Sea turtles can find their way through large areas of ocean.

Homing pigeons are good at finding their way back home.

Scientists think that some animals can feel Earth's magnetic field. These animals can use this feeling to help them know which way is north.

IN THE Real World

Lots of things that you use every day need magnets to work. This includes things that you might not expect. Magnets are in computers, earphones and even some trains.

Strong magnets are also found in big machines that doctors use to see inside someone's body. This helps doctors know if someone needs help in a part of their body they cannot normally see.

Be A Scientist

Now that you know more about magnets, you can try your own test to find out which materials in your house are magnetic and which are not. You will need a small magnet.

Never put a magnet near your face.

Collect some objects from around your house. Get a grown-up to check that it is okay for you to test each item with a magnet. Which objects does your magnet stick to?

Did anything surprise you?

23

Glossary

AFFECTED	to have been changed or touched by something
ATTRACTED	pulled toward something else
DIRECTION	the way that someone or something is moving or pointing
INVISIBLE	unable to be seen
MACHINES	things that help people do a task, which are sometimes powered by electricity or able to move in different ways
MATERIALS	things from which objects are made
METAL	a type of material that is heavy, shiny and sometimes magnetic
MODEL	a smaller version of something
OPPOSITE	different to or on the other side of something
TOOLS	equipment or instruments that are used to do a specific job

Index

ANIMALS 12, 18–19

COMPASSES 14–15, 18

EARTH, THE 12–15, 19

ENDS 10–11

FIELD 7–8, 13, 16–17, 19

FRIDGES 9

HOUSES 13, 22–23

IRON FILINGS 16–17

METAL 6, 9, 14, 16

NEEDLES 14–15

POLES 10–12, 14–15, 17

ROCK 6